# Alzheimer's, Autism and Music

*Maria Lavelle*

For Grandma...my best girl

And Oliver

# Contents

# Introduction

Cognition is greatly affected by those diagnosed with Autism and Alzheimer's despite modern, advanced medication and technology. This book assesses the value of music on both brain 'disorders' in order to establish not only a connection between the conditions themselves, but to discover if music in either a controlled or casual environment, displays any such evidence to suggest that music as a rehabilitative form could benefit a person's cognitive state.

***Alzheimer's, Autism and Music*** does not look to find a cure for the conditions through the implementation of music listening and interaction, but endeavours to highlight both positive and negative effects. It particularly focuses on rhythm as a musical element, and its ability to organise thoughts and movements in the case of autism, and recollection in Alzheimer's disease.

Cognitive improvement in one or both 'disorders' under a music-based therapy programme or during casual music listening are further assessed in terms of laws, theories, and practices. The absolutist approach (musical meaning lies in the musical structure alone) versus referentialist (experience of music occurs because of outside factors such as a lyrical presence, physical setting i.e theatre, stadium, nightclub, and one's emotional state at the time of musicking) debate is evident throughout the book. Where do Alzheimer's and autism symptoms, functioning's and responses sit in the dichotomy?

The following research has been facilitated with the assistance of literature and journals from musicologists, cognitive and neuroscientists, physiosocial and psychological professionals.

# Alzheimer's and Autism: Cause and Affect

Human brain development concerning chemical and cellular growth continues up until the age of five years (Understanding autism.com), and this is a particularly crucial period for the autistic brain.  In severe cases of autism parts of the cellular structure are lost, altered, and enlarged, resulting in abnormal stabilisation.

Sacks (2007) suggests that the abnormal neuronal migration, which occurs during the brain and nervous systems' development, migrate and settle incorrectly in the brains neural circuits.  Despite a reduction in the size of the cerebellum, the brain is generally larger than their counterparts as is hypothesised by Geschwind and Galaburda (Sacks, 2007), and could be caused by damage to immunological and immaturity of the left hemisphere during infancy. Contrastingly, The once 'ordinary' brain of the Alzheimer's patient begins a mass shrinking process (atrophy) due a loss of nerve cells (Peterson, 2002. pp 29).

The disease is from a collection of syndromes known as dementias, which all result in cognitive and behavioural problems.  It is possible that a regression in the Alzheimer's brain state is responsible for a 'shake-up' of the neuronal settlement, and could therefore be accountable for the production of anomalous right hemisphere dominance.  Although the cause(s) of Alzheimer's and autism are unknown, it is possible that genetic and environmental issues have an equal part to play in its development (NINDS).  The chances of developing the disease below the age of sixty-five are rare, although cases have been noted to affect people in their thirties, according to the National Institute of Neurological Disorders and Stroke (NINDS). The German physician, Alois Alzheimer, first explained Alzheimer's disease (AD) in 1906.  His findings displayed evidence showing the brains neurotransmitter, acetylcholine was extremely deficient in a person with this 'disorder' causing an uncharacteristically large number of proteins, namely amyloid and tau, to collect in the brain[1]. This reaction resulted in the forming of neurofibrillary 'tangles' and 'plaques' that disallow the channelling of neurons (chemical messages) and nutrients inside the nerve cells.  'As more neurons degenerate, more synapses-points of communication between cells-are destroyed' (Peterson, 2002. pp 29).

When the disease has saturated the entire brain, making contact with the brain stem, it becomes overwhelmed resulting in death.

Unlike Alzheimer's, autism does not result in death, and their memory recall is not affected.[2] Despite problems with cognition, those living

---

[1] In the 2006 Journal of Child Neurology (21:6: 444-449), Dr. Deborah K. Sokol et al discovered that the Beta-amyloid precursor protein (APP) found in the Alzheimer's brain was also evident at high levels in the severely autistic brain. 'Additionally, a complex relationship between age, acetycholinesterase, and plasma neuronal markers was found.'

[2] Jeanne A. Barton, a mother of an autistic child, and expert on the 'disorder' claims that through academic research, there are, to date, 'one-hundred and sixty mirror image parallels confirmed between autism and Alzheimer's' and '...are already close to one-hundred and forty parallels between autism and schizophrenia' (Autismhelpforyou.com: Autism... Alzheimer's... Schizophrenia - Comparison)

with autism tend to display high levels of intelligence.  It is possible that an increase of gray matter in the autistic brain, responsible for brain cognition, correlates with high level-level intelligence.  Although this is true of the 'normally' functioning brain, Dr. Manzar Ashtari of the children's Hospital of Philadelphia and Pennsylvania stated, '…In the autistic brain, increased gray matter does not correspond to high IQ, because this gray matter is not functioning properly' (Hug the Monkey: Gray and White Matters, 2007).

The intelligence levels of an Alzheimer's 'sufferer' could diminish as the disease makes its way through the brain, although during the earlier stages of the disease, a patients' struggle to recall forgotten information may signal that their original level of intelligence still exists.

Although the onset of Alzheimer's is a gradual process, while the disease continues to eat away at both the left *and* right brain hemispheres, the subtlety of the illness quickly lessons.  Initially, symptoms of the disease present themselves as difficulties with decision-making, problem solving, and short lapses in memory.  This progresses into a lack of awareness when social interaction, language comprehension, and an overall display and understanding of emotions are required.

Challenging behaviour is also characteristic of both 'disorders', and in the case of the autistic person, is the result of frustration at failure to organise thoughts, and a struggle to express emotion.  The aptitude of the human being to formulate new sentences, and match thoughts to words, takes place in the left hemisphere.  Both autistic and Alzheimer's 'sufferers' succeed in this function.  In autism, the damaged right hemisphere prevents the correct execution of these thoughts and words.  It is highly likely that the inability of the affected individual to express emotion is the reason for their lack of sociability.

An article presented at hugthemonkey.com, titled, *Mirror Neurons, Oxytocin, and Autism,* stated that gray matter was found to be at an increased level in the areas of the brain that dealt with sociability. The idea has been found by Dr. Ashtari's team to be, '...related to abnormal function of the mirror neuron system'.  Mirror neurons fire from nerve cells in the brain when we engage in activity, and when we observe others.  Therefore, gray matter in the areas of the brain that deal with social interaction could be the reason for an autistic person's lack of understanding.

Again, in the *Mirror Neurons, Oxytocin, and Autism* article, it referred to a study carried out by a PhD student, Jaime Pineda, and his colleague, Lindsay Oberman, from the University of California, San Diego.  Their initial studies found that 'the mirror neuron system is well developed by the time a child is seven years old'.  Further testing, with the use of electrocephalogram (EEG) monitoring was applied to a number of children suffering with autistic spectrum disorders (ASD). The overall result revealed that the mirror neuron system worked to a degree. Mirror neurons fired normally when the children watched video footage of their own family, but not of strangers. The findings demonstrated recognition of those who surround, and are a permanent and regular part of the autistic child's life. That which is unfamiliar is irrelevant.

In relation to Alzheimer's disease, J.D.E Gabrielli, author of the journal article 'Cognitive Neuroscience of Human Memory' (49; 1998) found that:

> In vivo metabolic imaging studies (e.g. Frackowiak et al 1981) and postmortem studies of late-stage AD patients (Brun & Englund 1981) find substantial damage to association neocortices in the frontal, parietal, and temporal lobes but relatively little compromise of primary visual, somatosensory, auditory, and motor cortices, the basal ganglia, or the cerebellum. The sparing of the basal ganglia and cerebellum may account for intact rotary-pursuit and mirror tracing in AD. The sparing of modality-specific cortices and the compromise of association cortices may account, respectively, for intact perceptual and impaired conceptual priming.

The finding indicates that the mirror-neuron system in the Alzheimer's brain is functioning to an extent, although the modality-specific cortices could be a contributing factor behind inability to recollect and process memories. The person with Alzheimer's disease, just as the autistic person does, eventually tends to respond, mainly, to objects and sounds as opposed to people or emotions. Could it be that in order for a persons mirror neuron system to function in terms of human observation and emotional reaction, the areas of the brain unaffected by the disease need to mirror the parts of the brain that have been affected in order to function correctly?

Autism is sometimes referred to as a disorder, although, is an autistic persons lack of ability to interact socially and emotionally with others, an ailment?  They can excel greatly in other areas of life, often surpassing the abilities of the 'ordinary' brained person.  Steven Pinker (1997) believes that 'mind-blindness' is used to describe the autistic perception of the world and its contents.  Their existence holds no emotional value or a basic understanding.  Everything is to be 'grasped with suitable mental machinery' (Pinker, 1997. pp 332).  Their lack of emotion allows them to achieve more, and 'in a sense, autistic children are right: the universe is nothing but matter in motion' (Pinker, 1997. pp 332).

'Mind-blindness' can be recognised as early as three years of age, and one experiment that demonstrates this is the 'false belief' task:

> The experiment lifts the rubber ducky out of the bathtub and puts it on the bed, takes a Polaroid snapshot, and then puts it back in the bathtub.  Normal three year olds believe that a photo will somehow show the duck in the tub.  Autistic children know it does not.
>
> (Pinker, 1997. pp 332)

The above example of the 'false-belief' test displays an advanced state of mind, but the other aspect of the autistic child is lack of emotional understanding.  An autistic child is unable to 'distinguish between someone looking in a box, and someone touching it' (Pinker, 1997. pp 332).  Yet, the 'Smarties test', which also assesses the 'false-belief' theory in those with autism, is the one that autistic children tend to fail.  The test involves a 'Smarties' tube, and its expected content versus actual content.  A child is shown that the tube does not hold chocolate, but, instead, a pencil.  They are told that a person outside of the room will attempt to guess the contents of the tube.  In the case of the 'normal' brain functioning child, they will say that the person will guess there is chocolate inside.  The autistic child will say that the person will guess there is a pencil in the tube.  Autism can only allow its 'sufferer' to view things as they are.  Their brain does not allow them to think about other possibilities.  They go straight for the result; the process of elimination is not an option.  They 'pass the test that is logically the same as the 'false-belief' task, but not about minds' (Pinker, 1997. pp 332).  Despite the recognisable trait of the autistic person: logical thinking, this is not always possible to put into practice when the organisational skills of the brain are interrupted or non-existent.  The above results support the notion that the autistic brain favours the absolutist view.

In the 'ordinary' brain, '... reflection and deliberation are involved in the completion of the response pattern...' yet, the autistic person does not use this technique in order to reach a conclusion (Meyer, 1956. pp 31).  They act on impulsion.  Meyer (1956, pp. 31) cited John Dewey (1934, pp. 59), 'Impulsion forever boosted on its forward way would run its course thoughtless, and dead to emotion...' Through the familiarisation of music via a therapeutic route, cognition could be assisted along, eventually allowing the development of social skills.  The law of affect relates to the previous statement as it supposes that the stalling of a response results in emotional outcome (Meyer, 1956).

The Alzheimer's experience of 'false belief' is not experienced nor explained in the same way.  In Alzheimer's it is rare for a 'sufferer' to fall victim to hallucinations, as acknowledged by Ronald Peterson (2002) in his book, *Mayo Clinic on Alzheimer's Disease*.

In autism, the logic plays the key part.  In Alzheimer's the imagination takes over, although this is normally a reaction to medication, and therefore, not a natural symptom (Peterson, 2002). It could be said that with Alzheimer's ability to unknowingly dismiss or create what is not there, they are more inclined to adopt a referentialist opinion.

Music is processed in the left hemisphere, and words in the right hemisphere (Storr, 1992). Interestingly, when music and words are working as one, as they do in songs, they are united in the right hemisphere (Storr, 1992. pp 38).  Could a lack of left hemispherical interference be reasoning behind the autistic persons inability to express their feelings through speech, and why their speech is delivered monotonously?  In particular, children with right hemispherical lesions can show that they possess excellent reading

---

[3] Contrastingly, those who are diagnosed with William's Syndrome (WMS) are known to display high levels of musical proficiency, and are able to emotionally interact with music. The comparative study of these syndromes allows for precise assessments in musical emotion to be undertaken, due to a sparing of musical ability regardless of their neurological disorders (Juslin & Sloboda, 2001). The main characteristics of the WMS 'sufferer' are high sociability, low IQ, experience and comprehension of musical emotions, and they are emotionally drawn to life, overall (Levitin, 2006). The WMS brain is about fifteen percent smaller than the 'average' brain, and researchers have tried to find whether an unusual development of the planum temporale (part of the temporal lobe) could be the reasoning behind WMS 'sufferers' perfect pitch, musical memory, and atypical music and language skills. (Sciencedaily.com).

Gloria lenhoff is a diagnosed WMS 'sufferer', and has an IQ of below sixty. Despite this, Gloria has a handle on thousands of arias in thirty-five languages, and can sing each one from memory, as well as being perfectly capable of interacting socially (Sacks, 2007).

skills from a young age, although it is obvious that they lack emotional understanding of the text.  Despite a musically autistic child's ability to display savant musical skills, s/he is incapable of interpreting the piece[3], due to an inability to communicate emotions (Storr, 1992. pp 37).

When a person is aware: has memories, feelings, physical capabilities, and is able to communicate them correctly, this is more simply referred to as a sense of 'self', which is explained and questioned by Sacks (2007, pp. 36).  However, does puberty, a time when the 'self' undergoes development, hold greater significance amongst the two 'disorders'?

Stephen Wiltshire, an autistic boy who is referenced by Sacks (2007, pp. 154), had '...erupted musical powers-huge powers' at the age of sixteen.  Besides possessing absolute pitch and the ability to recall complex chord structures, Stephen could automatically create and play melodies, which would last for several minutes, and with this could transpose the key of his compositions with ease.  Significant changes and functions occur in the brain of the autistic person from the womb to puberty, which allow such a talent, as Stephen's to develop overnight or manifest.  According to Understandingautism.com the brain forms a nervous system in the womb from the first trimester right through to three years of age.  From birth, our brains have developed the foundations for early survival; how we grow, change, and process information is a basic result of this formation.  'The developing brain is primed to make new neural connections and to prime away old ones that are not useful or accurate' (Levitin, 2006. pp 261), and this could be a contributing factor behind Stephen Wiltshire's sudden musicality.

Juslin and Sloboda (2001, pp 435-45) discuss 'Strong Experiences of Music' (SEM), informing us of the four emotional aspects: intense emotion, positive emotion, negative emotion, and mixtures of emotions/conflicting emotions.  We all come across at least one of these emotions when we hear a song or piece of music that, particularly, affects us on a deeper, emotional level (Juslin and Sloboda, 2001).  The emotion felt is independent from our friends and families' experience of the same musical piece, therefore, thus becoming apart of the 'self'.  It is possible that due to SEM being of a personal nature, particularly when we are teenagers: the Alzheimer's patient proves that we hold on to, and carry these emotions with us through life, even when every other aspect of our 'self' is 'lost'.  When they have seemingly lost all awareness of 'self', others and environment, music, because of its emotional content, enhances ones learning capabilities and recollection skills (Sacks, 2007).

The common denominator for the Alzheimer's 'sufferer's' past musical recollection is the time of life they recall it from: their teenage years (Sacks, 2007).  It is an interesting notion that a person who has little or no memory, who is unable to successfully interact and communicate with others as well as complete simple, everyday tasks, displays such an ability.  According to musicologist, Daniel Levitin (2006, pp 231), the reason for this experience is partly due to the fact that during our teenage years we are "emotionally charged", and embark on a journey of self-discovery.  This is a reason why music can be observed to play such a significant role in the case of the

Alzheimer's patient.  Nobody can truly change the way a person emotionally interacts with a song or piece of music, as 'music and musical preferences become a mark of personal and group identity, and of distinction' between the ages of fourteen and eighteen years (Levitin, 2006. pp 232).  Therefore, by the age of twenty, a person's musical preference is ascertained.  Juslin and Sloboda (2001, pp 423) quoted Tia DeNora (1999, pp 50), who stated, 'A sense of self is locatable in music, in that musical materials provide terms and templates for elaborating self identity'.  DeNora implies that regardless of a person's musical talent, a piece of music or a song will make an emotional link with us at one time or another during ones lifetime.  It is important to note that this is most likely to occur during puberty.

Through a positive response to such a rehabilitative practice, therapists along with close family relations to the Alzheimer's patient are analysed with results, implying that improvement with cognition is possible[4].

---

[4] An 'randomised controlled' experiment was conducted at a public psychiatric hospital in Rekem, Belgium by the department of Neurology at the University hospital of Gasthuisberg, Leuven, Belgium. An evaluation of the effects of music-based therapy on cognition in the demented brain was measured by the mini-mental state examination (MMSE) and the Amsterdam Dementia Screening test 6 (ADS 6). Behaviour was tested using the abbreviation Stockton Geriatric Rating Scale (BOP scale). It studied twenty-five diagnosed women over a period of three months, with assessments being made before the experiment, at six weeks, and finally, immediately following the study period. The results displayed a significant improvement with cognition in patients with moderate to severe dementia, although behaviour was not affected. This study would need to be repeated in order for the findings to be confirmed. Information on this experiment was presented at Sage Journals Online: Clinical Rehabilitation, Vole 18, No. 3, and 253-260. (2004).

The person diagnosed with Alzheimer's displays little evidence to suggest difficulty with the expression of emotion, whereas those diagnosed with autism find it challenging.  It is not that a person with autism finds it impossible to display basic emotion.  'Happy' and 'Sad' emotions are recognisable by the autistic person, and according to Heaton et al. (1999) as cited by Juslin and Sloboda (2001), being able to distinguish between the two during musicking is the extent of the autistic emotional capability.  This could be the reasoning behind an autistic person's display of savant musical ability.  Being able to differentiate between a 'happy' and 'sad' piece of music in autism according to Juslin and Sloboda (2001, pp 127), has been measured on the grounds of basic testing, in which the 'performance was generally poor'.  Yet, Heaton et al (1999), as cited by Juslin and Sloboda (2001, pp 127) stated, 'The only empirical study that has been undertaken in relation to musical emotion in autism has reported negative results'.  If a person can identify the basic emotions presented in a piece of music based on their innate feelings as opposed to outside influence, then is it not viable to suggest that that these emotions could be elaborated on until complete emotional understanding has been formed?  A basic connection with musical emotion shows that they are not merely 'robots' (Levitin, 2006. pp 259-60).  Instead of dealing with their thoughts on an emotional level, and then expressing them, the autistic 'sufferer' will act on them automatically.  This is a very logical way of dealing with, and processing life.  The information collected is constantly undergoing a scanning and sorting process.  This is the case with regular brain functioning (Storr, 1992. pp 37), although in autism this process is disrupted, and in severe cases, halted.  Music-based therapies could assist in the rehabilitation of the emotional and social problems faced by those diagnosed with autism.  Not all 'sufferers' of the disease display advanced musical ability or interest, which because of an unknown genetic abnormality creates neural dismorphology (Levitin, 2006).  Apart from neural dismorphology being the cause of social disinterest or musical talent, Levitin (2006) refers to geneticist, Julie Kronenburg, who theorises that there is a cluster of genes instrumental to musicality and outgoingness[5].

The majority of this chapter has looked at the differences and similarities of the two brain 'disorders' with an introduction to music, paying particular attention to the relevance of puberty.  The information collated has not just allowed for a basic understanding of the diseases, but through the determining of sociability, emotional understanding, and musical competency, has laid the foundations for ascertaining where the disorders sit in the absolutist versus the referentialist debate.  At this early stage of analysis the social barriers and the importance of logic over emotion as has been proven by the 'false belief' theory and mind-blindness test, suggesting that the autistic brain leans towards the absolutist school of thought.  One could be inclined to place the Alzheimer's sufferer in the category of the referentialist due to their need for social interaction, and their constant flow of emotion.

---

[5] **Those diagnosed with WMS mirror this theory, perfectly (Levitin, 2006).**

The following chapter will now elaborate on the classification of the two diseases whilst continuing to analyse the importance of music on cognition in a rehabilitative sense.

**Absolutist versus Referentialist: Music as Therapy**

According to the American Music Therapy Association, Inc (AMTA), music therapy provides a '...powerful and non-threatening medium', allowing patients to enjoy an individualised programme that assists the sufferer in the rehabilitation of their cognitive skills.
Sacks (2007), refers to and discusses the case study of an eighty-year-old Alzheimer's patient, Mr. Geist.  Although his memory had significantly deteriorated during his thirteen-year battle with the disease, he could remember music.  Mr. Geist had been a part of a singing group for forty years, and music was the only thing he could remember through his illness, and in detail.  He managed to open a large musical performance whilst 'suffering' from the disease, but '...the evening he performed, he had no idea how to tie a tie...he got lost on the way to the stage-but the performance?  Perfect...  He performed beautifully and remembered all the parts and words' (Sacks, 2007. pp 339).  Mr. Geist had previous musical experience; therefore the music was already embedded in his memory.  Due to this, the preceding example could be viewed as a biased one.  It could be said that music may not necessarily have the same effect on a sufferer who lacked in musical knowledge or experience.  Meyer opposes the idea that music opens the doors to our thoughts and emotions.  He believes that 'without thought and memory there could

be no musical experience' (Meyer, 1956. pp 87).  This theory links to Meyer's absolutist approach to musical emotion.  It suggests that there is never a complete loss of memory in Alzheimer's: it just struggles, and at times fails to surface.  The brain undergoes numerous reactions, which in turn enable us to organise our thoughts and emotions.  Speaking, body language, and writing are all forms of communication, and they function as organisational tools that help to present outside of ourselves that which we have organised inside the brain.  In accordance with Sacks (2007), music allows us to embed sequences that assist us in thought arrangement and recollection.  A prime example of this is the A, B, C... and the 1,2,3...songs, as referred to by Levitin (2006), which are used to help children learn and easily store away the information in the memory bank.  According to Thaut (2008), it is because of the rhythmic structure that this can be achieved, although Sacks (2007) referred to rhyme, song, and meter as the most powerful mnemonic devices.  He believes mnemonic devices are utilised by all types of people whether to learn rules, the periodic table, the order of notes on the musical stave, or to revise for an exam.  In the case of the next example, melody was the key to his memory recall.  The composer Ernst Toch could recall a complete set of numbers after hearing the order just once, as is acknowledged by Sacks (2007).  Through the creation of a melody, Toch was able to pair the numbers with the notes.  Due to this

---

[6] Whereas music has the ability to unlock the bodily restraints of a Parkinson's patient, music and lyrics are required in the rehabilitation of the aphasia sufferer, as is stated by Sacks (2007. pp 336).

association Toch could memorise any numerical sequence by relating a number to a musical note[6].

The Law of Pragnanz states that what memory cannot figure out correctly, or decipher, will be forgotten.  It 'functions within the memory process, which tends to complete what was incomplete, to regularise what was irregular, and so forth' (Meyer, 1956. pp 89). This is indicative of the Alzheimer's patient's ability to subliminally decide upon the types of memories that they recall, and is based on the simplicity of the contributing factor, which enables a particular memory to be created.  The ease of a memory formation equals the ease of memory selection, resulting in why rhythmic and melodic phrasing plays a significant role in memory selection.  The combination of these two musical elements allows 'chunking' to come into play: a mechanism, which assists in our declarative learning and recall, and motor learning.

Thaut (2008) refers to research, which '... is showing that music can influence perceptional organisation of behaviour in autistic children', and therefore, Sue Bettison's report, which was released in 1996, stating that cognitive measures in autistic children can be greatly benefitted through rhythmic music listening, and can begin to be taken more seriously.  Bettison's theory had previously been rendered inconclusive due to '...the absence of a control group...' in her study (Thaut, 2008. pp 78).  As well as Thaut's scientific findings, other studies have been conducted in an attempt to find a relationship between music and cognition in both the 'normal' and 'disordered' brain.  Heaton et al (1999), and Ma et all (2001) are two researchers who have similarly found that '...executive functions, memory, and affective reasoning, as well as psychological functioning...' have been improved through the implementation of musical exercises (Thaut, 2008. pp 78).

Referring back to the A, B, C... and 1,2,3... songs, which are prime examples of how the 'chunking' process can be simply understood, the melodies and lyrics of these songs are learnt parallel to the rhythm.  Rhythm helps one to organise movement, speech, thoughts, and memories (Thaut, 2008. pp 78-9).  Thaut referred to Cowles et al. (2003) when stating that patients with Alzheimer's disease retained musical memory longer than non-musical memory. According to Thaut (2008, pp 75) '...music can serve as an effective mnemonic device to facilitate verbal learning and recall in healthy persons, patients with memory disorders, and children with learning disabilities'.

Our short-term memory, which is stored in the pre-frontal cortex, is measured in 'chunks' rather than sounds.  Through the 'chunking' of sounds we are able to break down what we hear, and label them.  It acts just as a filing system, and by labelling our short-term memory chunks we are then able to store them away in our long-term memory bank until we need or want to recall them.  It is an innate attribute that is spread across a broad range of nervous systems (Thuat, 2008. pp. 75).  Alzheimer's attacks the hippocampus where our long-term memory is stored, and is reasoning for why memory function could be induced by the information presented through a musical context.  Music is an extremely important emotional force (Thaut, 2008).

One of Sacks' (2007) 'tales' refers to a six-year-old autistic boy named David.  The task of tying his laces was impossible, yet his audio-motor-co-ordination was second-to-none.  When a therapist got David to tie his shoelaces to a song, he achieved it on the second attempt.  David may have always understood the process behind tying his shoelaces, but lacked the timing required to structure each part of the process.  Rhythm is timed, and David was able to structure his thoughts thus completing the task successfully.  Some may argue that it is down to the melody alone, or the timbre, or harmony, which make it accountable for the organisation of thoughts and speech.  Thaut states that melody would not exist if it lacked the rhythm/timing needed to structure and piece of music.  According to Storr (1992, pp 33) melody and harmony do not have the same effect because 'rhythm is rooted in the body in a way which does not apply so strikingly to melody and harmony'.  Rhythm allows 'chunking' to take place in the brain thus rendering it probable that rhythm is what unites language and music in the brain.  'To study rhythm is to study all music' (Storr, 1992. pp 33).  The above statement suggests that rhythm could be responsible for neural entrainment.  Thaut (2008) hypothesised that rhythm and music assisted in the creation of a sensory structure through temporal ordering, which allowed the autistic child to map his attention.  It was reported by Thaut and Mahraun (2004), as cited by Thaut (2008, pp 77), that there is 'preliminary evidence for the temporal entrainment of attention in autistic children'.  Scientific data and research carried out by Hermelin et al. (1989), Thaut (1998), Wimpory et al. (1995), Heaton et al. (1999), and Ma et al. (2001) has provided evidence that show musical exercises have a positive effect on 'executive functions, memory, and affective reasoning as well as physiosocial functions' (Thaut, 2008. pp 78).  This occurs due to music's ability to 'access control processes in the brain related to control of movement, attention, speech, production, learning, and memory, which can help retrain and recover functions in the injured or diseased brain' (Thaut, 2008. pp 116).

Interestingly, Pinker, who disagrees with the idea that music has an affect on cognition in any way, states, 'Rhythm is the universal component of music, and in many idioms it is the primary or only component' (Pinker, 1997. pp 539).

Fitt's Law is a psychological theory that suggests the faster a task is undertaken; the number of errors is likely to increase.  This law does not apply to the autistic person.  Thaut (2008) reports that as the speed of the task increased, there was a significant reduction in the number of mistakes made.  The interest in this finding was heightened when it was realised that music and rhythm played a major part in the speed and accuracy of the task.  One study undertaken involved a group of subjects, half of whom were autistic. This particular sustained attention task used card matching as its method of research.  The results showed that the autistic individuals performed much better when listening to music and rhythm patterns (Thaut, 2008).  Additionally, from the above study and similar ones that have taken place over recent times, it has been established that 'a faster task completion equalled decrease of inappropriate behaviour' (Thaut, 2008. pp 77).  Therefore, the implementation of music and rhythm improved attention in all areas of the autistic child and their behaviour.

Apart from the importance and relevance of the musical elements, the way in which a person listens to music could prove significant. Psychologist, Kurt Koffka, theorised the 'averaging effect', which is presented by Meyer (1956).  When we hear something, we fail to store it exactly as it is.  According to Koffka, even the healthiest brain possesses an unstable memory trace.  This suggests that in the case of the Alzheimer's patient, the 'averaging effect' is particularly prominent, and as a result the sufferer is unable to truly experience a piece of music when it is recalled.  It could be suggested that only when an original recording or a live performance of the piece is performed that a patient is able to emotionally connect with the music or song.

A colleague of Koffka's, and co-founder of the Gestalt school of Psychology, Max Wernheimer, produced an idea: the 'thought process' theory, which is presented by Meyer (1956, pp 87-88):

'....S1......S2.....'

Wernheimer proposes the idea that 'two situations' take place during the thought process: the first situation (S1) followed by the second situation (S2).  Wernheimer believes that S1 does not just occur.  It requires a root, or a starting point.  In order to recall a memory, or to conceive a thought, something has to occur visually, or we have to hear something: a sound, lyrics, gossip, or we need to have physically interacted with a person or object in order to kick-start the thought process.  Meyer (1956, pp. 88) states, 'The relationship between structural troubles and their resolutions are intelligible and

27

resolvable'.  Therefore, the same theory could be applied to both the Alzheimer's and autistic person's thought process.  It demonstrates the brains continual fight in terms of memory and awareness.  When the completion of the second situation has been achieved, 'the very solution does not represent an end but that by its very nature leads to further dynamic consequences' (Meyer, 1956. pp 88).  Arguably, in the case of the Alzheimer's 'sufferer' who has reached the severe stages of the disease, Wernheimer's thought process theory should not work.  This is due to the progression of the disease to the pre-frontal cortex, resulting in the loss of short-term memory.

The question is, can Wernheimer's theory only be truly effective during musicking?  Rigorous testing would have to be undertaken in order to establish if, and which musical element is responsible for a completion of the process.  If Wernheimer's theory is correct, then music listening does not stimulate the brains thought process, or memory recall, but is merely working as a catalyst.  By reaching a conclusion of some form in the cases of Alzheimer's and autism, it could provide hope in knowing that the affected brain is striving for a resolution.  Despite both the Alzheimer's and autistic 'sufferers' struggle to seek a resolution in all they do and think, they separate on the absolutist versus the referentialist dichotomy.

The autistic person sees all things logically thus viewing life from a completely different perspective.  To a child with autism, 'cuddly dolls and stuffed animals hold little interest', and this is also true when surrounded by other children (Pinker, 1997. pp 331).[7]

---

[7] In the case of the popular autistic savant, Kim Peek, who is more commonly referred to as the 'Rain Man', and those who live with a similar form of autism, they possess the ability to '...learn

Despite autism viewed by some as an ailment because '...they touch, smell, and walk over people as if they were furniture', or because they play with a hand as though it were a mechanical object, those who adopt the idea do not shy away from the fact that '...the intellectual and perceptual abilities of some autistic children are legendary' (Pinker, 1997. pp 331).  A display of profound ability/talent in one or many areas of the autistic child's life will usually appear before the age of ten, and this particularly true of the musically gifted autistic child, according to Sacks (2007).  This observation relates to a quote by Francis Xavier, the first Jesuit, cited by Morag Stlyes (1989), which stated, 'Give me a child until he is seven and I'll give you the man'.  When research carried out by Dr. Fred R. Volkmar, an expert in autism, and Director of the Centre at the Yale School of Medicine, provides evidence to suggest otherwise, can Sacks and Xavier's claims be supported.  Jane Lytle is a parent of a nineteen-year old autistic boy who showed improvement over time, and still continues to do so.  She speaks of the significant and positive changes throughout his life so far, and in an interview for the Washington Post, titled *Raising an Autistic Child*, Dr. Volkmar stated, 'By six he had turned a significant corner'.  It must be noted that music based therapies were not the reasons for these improvements, although the examples do supply evidence that suggests improvements can be made, especially during the teenage years.

---

multiplication at rapid speeds, put together jigsaw puzzles, disassemble appliances, read distant license plates, or instantly calculate the day of the week on which any given date in the past, or future falls'" (Pinker, 1997. pp 332),

It could be suggested that the musically inclined autistic person is generally viewed as the one who displays profound musical talent. The musical performance of the 'musical savant' has been qualified at times as 'mechanical' (Mottron et al. 1999; Sloboda et al. 1985).  This could be due to the autistic characteristic that solely draws the person to the structure.  An emotional detachment may allow for a rational musical performance, permitting an audience to experience the intended composition, and not a personalised piece.  This theory relates to the idea of the absolutist view, presented by Meyer (1956, pp 33), which explains, 'the meaning of music lies specifically, and some would assert, exclusively, in the musical processes themselves'. The amygdala, and other medial parts of the brain may be poorly developed in people with aspergers, as is stated by Sacks (2007)[8]. Musical interest and ability go hand-in-hand with competent social skills, although, surprisingly, this can mean the complete opposite with some autistic people.  As this could lead one to consider the effect of music therapy altogether, the relevance of music to the autistic person could be viewed as a key factor in the development of cognition.

---

[8] Temple Grandin, both a neuroscientist and friend of Sacks, is a 'sufferer' of autism, and whilst attending a performance of Bach's 'Two and three-part inventions', Temple revealed that she 'got intellectual pleasure from Bach, but nothing more' (Sacks, 2006. pp 290). Additionally, whilst driving through the mountains with Temple one day, she displayed a profound lack of emotion. Sacks found the sights emotionally stimulating, yet, Temple commented that 'the mountains are pretty, but they don't give me that special feeling' (Sacks, 2006. pp 291). Again, this could be viewed as a weak form of study, and therefore, quantitative research may be the only way to record accurate findings.

Individually tailored and group-based music therapies could play a significant role in the rehabilitation of a patient.  Can group-based music therapy benefit the socially inept autistic 'sufferer' with their cognitive skills, and, similarly, would an individualised music-based therapy accommodate the Alzheimer's patient, resulting in more positive outcomes with cognition?  An idea put forward by DeNora (1999), which was presented by Juslin and Sloboda (2001, pp 423), proposed that music acts as a 'mirror', allowing a person to see oneself.  This could suggest that only individualised therapies would produce the greatest results, as each person's reaction would be unique to their own musical and emotional experience.

Music therapist of more than ten years, Gretta Sculthorp, put into words the effect music therapy had on her patients.  She wrote:

> ...There are listeners who come and stand beside or in front of me, touching me for the whole time.  There are always people who cry.  There are people who join in-for operetta or for Sinatra songs (and Leider, in German!).  There are disturbed people who don't know who they are, but recognise me immediately as "the singing lady".  (Sacks, 2007. pp 345)

Even though her patients did not display the same responses across the board, it is evident that each different reaction was expressing various thoughts and emotions.  If we break down the experiences of each reaction, we can gather information on their emotional state.  For those who come and stand alongside her, touching her for the whole time, it could suggest that they want to get as close as they can to the music.  The everyday battle they go through in their mind, to find a place or moment they recognise, is so close when music, or singing can be heard: they don't want to let go of the possibility of reuniting with their 'self'.  Patients may cry because they are possibly aware that the feeling will not last for very long.

In the early 1960's, Paul Nordoff and Clive Robbins were the first to conduct pilot projects in the utilisation of music therapy with severely autistic children.  According to the American music association, Inc (AMTA), 'Music is a very basic human response'.  Those who are diagnosed with autism and the various other disorders that fall under the umbrella of ASD, have innate musical interest and ability.  Therefore, a sense of security, which is vital to those who are diagnosed, can be established through music therapy.  Autism requires familiarity, repetition, and resolution, which music therapy offers, and could particularly benefit from an individualised programme.  It promotes relaxation that can unlock the 'restricted repetitive and stereotyped patterns of behaviours, interests, and activities'.  (AMTA) [9]

From the information presented in this chapter, it could be proposed that the Alzheimer's sufferer would benefit from a regular group-based music therapy.  This in itself would classify Alzheimer's as a referentialist disease, and autism an absolutist syndrome.

The following section will now look deeper into the affects of music and music-based therapies on the brains of the two 'disorders', thus establishing a concise and impartial conclusion.

------

[9] Sacks (2007, pp 252) provides an example of how music therapy was able to replace a Parkinson's sufferers '...compressed, clenched, blocked, or else jerking...' frame with '...blissful ease and flow of movement...'

## Conclusion

This concise study set out to analyse the effects of music in a therapeutic sense, on the Alzheimer's and autistic brain. It covered theories and methodologies, of which the majority could be related to both 'disorders'.

It has been proven that Alzheimer's patients are able to retain musical memory longer than non-musical memory (Thaut, 2008), although any form of music therapy in the latter stages of the disease could only assist in the *easing* of its symptoms. Such therapies would be constantly battling against Alzheimer's, meaning that the results would be short-lived. This is why an individually tailored programme in a family-grouped environment may best suit the 'sufferer'.

The autistic brain lacks a fully functioning mirror neuron system, which deals with social and emotional awareness. From this observation it could be proposed that a creation of a music programme, based on the stalling process in relation to the law of affect (Meyer, 1956), may assist in the rehabilitation of emotional expression and sociability. Initially, it could make way for challenging behaviour as the autistic person's strive for resolution would be interrupted, although this would be part of the developmental process.

As we have found out, the averaging affect does not apply to the autistic brain, and because 'sufferers' perfectly store away what they witness, experience, and hear, there is a possibility that they could eventually learn to experience and understand emotion that goes further than a 'happy' and 'sad' differentiation.  If such a curriculum were developed, the patient would initially benefit from an individually tailored therapy where emotions and behaviour would be addressed and improved.  Eventually, the autistic 'sufferer' could be integrated into a group situation where sociability would be enhanced.

Having placed the 'ailments' into suitable rehabilitative environments, the implementation of musical practice in terms of an absolutist and referentialist approach needs to be decided.  In this investigation, Alzheimer's was quickly labelled as a disease in favour of the referentialist view.  At this stage of research, the findings within this book suggest it remains the same.  The type of music required need not be too deliberated.  As it was presented in chapter one, music from the Alzheimer's teenage years was of particular importance, and as the disease is of a progressive nature, the therapy is more than likely going to succeed from the utilisation of personal musical pieces. The autistic approach to musical preference in a therapeutic setting could be seen as slightly more complex, as the 'sufferer' has time to improve.

Rhythm has been said to boost neural entrainment as well as memory function, and for those reasons, the use of rhythm could act as a mnemonic device in the learning and improvement of cognition. Rhythm as a musical element could be positioned under the heading of the absolutist: the logical approach to learning.  Alternatively, it is possible that through the application of melodic phrasing, tone, and harmonics, the autistic person could begin to experience emotion: the referentialist method.

It is clear from the research that the two 'disorders' originally stood firmly at opposite ends of the absolutist versus referentialist debate. As I reach a conclusion I become aware of an autistic brain's ability to progress: acquiring varied levels of cognition, thus possibly making way for the development of a referentialist train of thought. Throughout this whole project, Wernheimer's 'thought process' theory, if ever validated by the scientific world, could hold the most important information in terms of rehabilitation as a whole.  If music is only functioning as a catalyst, the brain must already be working alone, to some degree, providing hope that the 'afflicted' individual's brain is still aware.

Music has the ability to induce certain emotions in some, and promote neural entrainment in others, and although they do not provide cures for some diseases, if any, what they do promote is a non-invasive alternative, or compliment to medication and/or treatment.

The study of musical affect in cognitive science is relatively new. Therefore, it could be likely that an in depth investigation into the relationship and relevance of absolutist and referentialist grouping in terms Alzheimer's and autism has not been embarked upon.  With this in mind, it is possible that a well-developed study of this nature could influence future scientific studies in the understanding of Alzheimer's, autism, and other brain disorders.  Through an establishment of where a particular disease leans in terms of the absolutist and referentialist schools of thought, it is possible that a specifically tailored music therapy programme can be designed to assist in rehabilitation.

The research and analytical study that assisted in the development of this project has proved my hypothesis: music does have an affect on the Alzheimer's and autistic brain.  Indeed, it must be reiterated that music does not always produce positive results with reference to musicogenic epilepsy, although there will always be a patient who physically and/or mentally disagrees with a form of rehabilitation or type of medication.

Based on research of this size it would be bold and highly inflative to suggest that every area of investigation has been utilised.  The accumulation of information from texts: both anecdotal and scientific, provided this study with sufficient knowledge on this subject, allowing for the development of my own theory(ies).

Without first hand experience in a controlled environment, I do not believe that the 'ailments' and their relationship with music can be fully understood.  Therefore, as a further line of enquiry I propose a combination of larger practical examinations within a scientific environment.

## Bibliography

American Music Therapy Association, Inc. 2004.  Available at:
http://www.musictherapy.org/factsheets/MT%20Autism%202006.pdf
[Accessed 21 November 2008]


Autism… Alzheimer's… Schizophrenia- Comparison. 2002. Is all this "just coincidence"? [Online] (Updated 2003) Available at:
**http://www.autismhelpforyou.com/Autism%20Alzheimer's%20Schi zophrenia%20Compared.htm**
[Accessed 18 January 2009]


Bettison, S., 1996. The long-term effects of auditory training on children with autism. Journal of Autism and Developmental Disorders, 26, 361-375


Brun A, Englund E. 1981. Regional pattern of degeneration in Alzheimer's disease: neuronal loss and histopathological grading. Histopathology 5. p 549-64


Cowles, A., Beatty, W. W., Nixon, S.J., et al. 2003. Musical Skill In Dementia: A Violinist Presumed to Have Alzheimer's Disease learns to play a new song. *Neurocase*, 9, p 493-503


DeNora, T., 1999. 'Music as a technology of the self' *Poetics,* 27 (1), p 50.


Dewey, J., 1934. Art as experience. USA: Perigree Books


Frackowiak RS J, Pozzilli C, Legg NJ, Du Boulay GH, Marshall J, et al. 1981. Regional cerebral oxygen supply and utilization in dementia: a clinical and physiological study with oxygen-15 and positron tomography. *Brain* 104, p 753-78


Gabrieli, J. 1998. Cognitive Neuroscience of Human Memory. [Pdf] p, 49. Available at: http://web.mit.edu/gabrieli-lab/Publications/1998/Gabrieli.Ann.Rev.psych.1998.pdf  [Accessed 22 March 2009]

Heaton, P., Hermelin, B., and Pring, L. 1999. Can Children with autistic spectrum disorders perceive affect in music? An experimental investigation. Psychological medicine, 29, p 1405-1410.


Hermelin, B., O'Connor, N., Lee, S., et al. 1989. Intelligence and musical improvisation. Psychological medicine, 19 (2), p 447-457.


Hug the Monkey. 2007. Autism: Gray and White Matters. [Online] (Last updated 27 November 2007) Available at: http://www.hugthemonkey.com/autism/ [Accessed: 2 February 2009]


Hug the Monkey. 2007. Autism: Mirror Neurons, Oxytocin, and Autism. [Online] (Last updated 9 November 2007) Available at: http://www.hugthemonkey.com/autism/ [Accessed 22 March 2009]


Juslin, N. P., Sloboda, A. J., 2001. Music and Meaning: Theory and Research. New York. USA: Oxford University Press


Hyman, M. 2009. Autism: The Role of Inflammation and the Example of Autism. [Online] Available at: http://www.ultramind.com/epidemic_autism.php [Accessed 1 April 2009]

Lindstrom, E., 1997. Impact of Melodic Structure on Emotional Expression: In Proceedings of the Third Triennial ESCOM Conference, Uppsala, (ed. A. Gabrielsson), pg. 292-7. Uppsala, Sweden: Uppsala University


Levitin, D., 2006. This is Your Brain on Music: Understanding a Human Obsession. Great Britain: Atlantic Books


Lytel, J., Volkmar, F. 2008. Raising an Autistic Child. The Washington Post. [Internet] (Last updated 18 November 2008) Available at: http://www.washingtonpost.com/wp-dyn/content/discussion/2008/11/14/DI2008111402395.html [Accessed 9 February 2009]

Ma, Y., Nagler, J., Lee, M., et al. 2001. Impact of music therapy on the communication skills of toddlers with pervasive developmental disorder. Annals of the New York academy of sciences, 930, p 445-447.

Meyer, B. L., 1956. Emotion and Meaning in Music. Chicago. USA: The University of Chicago Press

Mottron, L., et al. 1999. Absolute Pitch in Autism: A case-study. *Neurocase,* 5, p 485-501

National Institute of Neurological Disorders and Stroke: National Institutes of Health. (Updated 27 March 2002) [Online] Available at: http://www.ninds.nih.gov/disorders/alzheimersdisease.htm?css=print [Accessed 20 December 2008]

Peterson, R., 2002. Mayo Clinic on Alzheimer's Disease. [E-book] Philadelphia: Mason Crest. Avaask in children with autism. Submitted for publication.

Thaut, M.H., 1998. Measuring musical responses in autistic children: A comparative analysis of improvised musical tone sequences of autistic, normal, and mentally retarded individuals, *Journal of autism and developmental disorders*, 18, p 561-571

Understandingautism.org: Threshold. 2000-03. Fact-based Autism Information: Biological Developmental Domain. [Online] Available at: http://understandingautism.com/ [Accessed 29 January 2009]

Van De Winckel, A., Feys, H., De Weerdt, W. 2004. Clinical Rehabilitation. *Cognitive and behavioural effect of music-based exercises in patients with dementia.* [Online]. 18 (3), p 253-260. Abstract from Sage Journals online. Available at: http://cre.sagepub.com/content/abstract/18/3/253 [Accessed 2 April 2009]

Wimpory, D., hadwick, P., and Nash, S. 1995. Brief report: Music interaction therapy for children with autism, an evaluative case study with two-year follow-up. Journal of autism and developmental disabilities, 25 (3), p 231-248.


http://www.ninds.nih.gov/disorders/alzheimersdisease/alzheimersdisease.htm  Last updated 27th March 2009 National institute of neurological disorders and stroke. Alzheimer's disease information page. Accessed 13 December 2009


ilable at:
http://www.questia.com/read/114493449?title=mayo%20clinic%20on%20alzheimer's%20disease [Accessed 2 April 2009]


Pinker, S., 1997. How the Mind Works. UK: Penguin


Sacks, O., 2007. Musicophilia: Tales of Music and the Brain. Great Britain: Picador


Scientific commons Beta. 2002. Neuronal Migration and Neuronal Migration Disorder in Cerebral Cortex. [Online] Environmental medicine: annual report of the research institute of environmental medicine, Nagoya University. Available at: http://en.scientificcommons.org/8918182 [Accessed 7 April 2009]


Science Daily. 2008. Autism and Schizophrenia Share Common Origin, Review Suggests. [Online] (Last updated 18 December 2008)
Available at:
**http://www.sciencedaily.com/releases/2008/12/081216114746.htm**
[Accessed 20 January 2009]


Sloboda, J. A., et al. 1985. An exceptional musical memory. *Music perception,* 3, p 155-70


Sokol, KD, PhD, MD et al. 2006. High Levels of Alzheimer Beta-Amyloid Precursor Protein (APP) in Children With Severely Autistic Behaviour and Aggression. *Journal of Child Neurology*, [online]. 21, (6), p 444-440. Abstract from Sage Journals Online.

Available at: http://jcn.sagepub.com/cgi/content/abstract/21/6/444 Accessed on: 31 March 2009.


Styles, M., 1989. Collaboration and Writing. USA: Open University Press.


Thaut, H. M., 2008. Rhythm, Music And The Brain: Scientific Foundations and Clinical Applications. New York. USA: Routledge


Thaut, M.H., and Mahraun, D., 2004. The influence of music and rhythm on a sustained attention t

# A Special Acknowledgement

I would like to thank you, Dr. Benjamin MacPherson for all of your advice and support throughout and towards this project.  Thank you very much.

Made in the USA
Monee, IL
07 July 2026

56550174R00028